I0045466

Thomas G. Rooper

A Pot of Green Feathers

A study in apperception

Thomas G. Rooper

A Pot of Green Feathers
A study in apperception

ISBN/EAN: 9783337248192

Printed in Europe, USA, Canada, Australia, Japan

Cover: Foto ©berggeist007 / pixelio.de

More available books at **www.hansebooks.com**

Teachers' Professional Library.

A Pot of Green Feathers."

A STUDY IN APPERCEPTION.

DISCUSSION OF THE MENTAL OPERATIONS BY WHICH
WE ACQUIRE KNOWLEDGE.

BY

T. G. ROOPER, M.A.,

INSPECTOR OF SCHOOLS IN ENGLAND.

NEW YORK AND CHICAGO:

E. L. KELLOGG & CO.

Published by special permission of the author. This edition has been carefully compared with the original, and is the only correct one on the market at this time.

COPYRIGHT, 1892, BY

E. L. KELLOGG & CO.,

NEW YORK.

———

'A POT OF GREEN FEATHERS."

CONTENTS.

3

VALUABLE TEACHERS' BOOKS

AT LOW PRICES.

Reading Circle Library Series.

Allen's Mind Studies for Young Teachers.
"   Temperament in Education.
Welch's Talks on Psychology.
Hughes' Mistakes in Teaching.
"   Securing Attention.
Dewey's How to Teach Manners.
Woodhull's Easy Experiments in Science.
Calkins' Ear and Voice Training.
Browning's Educational Theories.
Autobiography of Froebel.

Bound in cloth. Price 50 cents each; to teachers, 40 cents; by mail, 5 cents extra.

Teachers' Professional Library Series.

Reinhart's History of Education.
"   Principles of Education.
"   Civics of Education.
Blackie's Self-culture.
Browning's Aspects of Education.

Limp cloth. Price 25 cents each; to teachers, 20 cents; by mail, 3 cents extra.

Teachers' Manual Series.

19. Allen's Historic Outlines of Education.
18. Kellogg's The Writing of Compositions.
17. Lang's Comenius.
16. " Basedow.
15. Kellogg's Pestalozzi.
14. Carter's Artificial Production of Stupidity in Schools.
13. McMurry's How to Conduct the Recitation.
12. Groff's School Hygiene.
11. Butler's Argument for Manual Training.
10. Hoffmann's Kindergarten Gifts.
9. Quick's How to Train the Memory.
8. Hughes' How to Keep Order.
7. Huntington's Unconscious Tuition.
6. Gladstone's Object-teaching.
5. Fitch's Improvement in the Art of Teaching.
4. Yonge's Practical Work in School.
3. Sidgwick's On Stimulus in School.
2. Fitch's Art of Securing Attention.
1. Fitch's Art of Questioning.

Price 15 cents each; to teachers, 12 cents; by mail, 1 cent extra.

Our large descriptive catalogue of valuable books on teaching free on application.

E. L. KELLOGG & CO., New York and Chicago.

INTRODUCTION.

As the title of this paper seems a little strange, a few words are necessary to explain its meaning. Some years ago I was listening to an object-lesson given to a class of very young children by a pupil-teacher who chose for her subject a pot of beautiful fresh green ferns. She began by holding up the plant before the class and asking whether any child could say what it was. At first no child answered, but presently a little girl said, " It is a pot of green feathers." Thereupon the teacher turned to me and said, " Poor little thing! She knows no better."

But I fell a-thinking on the matter. Did the child really suppose that the ferns were feathers ? Or did she rather use the name of a familiar thing to describe what she knew to be different, and yet noticed to be in some respects like ? This train of thought led me to put together what I knew of perception, and the following is the result of my labors.

The principal authority which I have closely followed is Dr. Karl Lange's Apperzeption, but I have derived much help from Herbart's Psychology, Bernard Perez's " First Three Years of Childhood," Romane's Mental Evolution in Man, and the lectures of the late Professor T. H. Green.

5

"THE POT OF GREEN FEATHERS."

I.

Some Questions.—What do we know of the outer world? Of that which is not self? Of objects? How do we know *anything* of the outer world? We receive impressions from it ; a table feels hard, a book looks brown in color, oblong in shape, and we say it is thick or thin. Are we simply receivers of these impressions —hard, brown, oblong? Are our minds inactive in the process of getting to know these impressions? Or are they active? Are lumps of the outside somehow forced in upon our minds entire without corresponding action on the minds' part ?

What the Mind Does.—No ! our minds are not passive; the opposite is true. Through the senses the mind receives impressions, but these contributions from the senses would not be objects of knowledge, would not be interpreted, would not be recognized unless the mind itself worked upon them and assimilated them, converting the unknown stimulus from without into a sensation

7

which we can hold in our thoughts and compare with other sensations within us. The mind converts the unknown stimulus from without into the known sensation. The outer world then is no more wholly the outer world when we know it. In our knowledge of the outer world there is always something contributed by the mind itself.

Different Minds receive Impressions Differently.—The truth that the mind adds to and changes the impressions which it receives through the senses is illustrated by the very different conceptions which exactly the same landscape gives rise to in different people. The geologist can tell you of the strata, the botanist of the vegetation, the landscape painter of the light and shade, the various coloring, and the grouping of the objects, and yet, perhaps, no one of them notices exactly what the others notice. A plank of wood, again, seems a simple object, and able to tell one tale to all, but how much it tells to a joiner, concerning which it is dumb to a casual observer.

As we Advance Impressions affect us Differently.—Or again, visit as a grown man the school-room, or playground where you played as a child, especially if you have not visited the scene in the interval. How changed all seems ! The rooms that used to look so large have become dwarfed. The tremendous long throw which you used to make with a ball from one end of the playground to the other, to what a narrow distance it has shrunk ! Yet the room and the ground are what they were. It is your mind that has changed. The change in your mind has brought about for you a change in the thing. Two people, then, or even one's self at different times of one's life, may perceive the same object without obtaining the same perception. Yet if the external object stamped itself on the mind as a seal or die stamps itself on wax, if the mind were as passive as wax, how could one object give rise to such different impressions? The difference must be due to the mind.

New Impressions referred to Previous Ones.—Neither is it difficult to understand that this is so if we think what is the nature of the process by which the mind interprets the impressions which it receives from outward objects. When the mind receives an impression it refers it to a previously received impression that happens to resemble it. Thus every new impression is interpreted by means of old ones, and consequently every new perception is affected, colored as it were, by the already acquired contents of the mind, and nothing can be known or recognized at all until reference and comparison have been made to previous perception.

II.

Explanation of Perception.—My object to-day is to make this point, *perception*, which I admit is not easy, as clear as I can make it. Perception of an object is not so simple a matter as at first it seems to be. "Oh," some one will say, "simple enough! A dog runs by me; through my senses I receive sensations of the animal, and I know that I see a dog." But this is no perfect account; for suppose two strange animals, say, a *Tasmanian Devil* and an *Ornithorhynchus,* come up the street together, my senses will make me aware of their presence, but if I have not learned anything about them previously, I shall not know, I do not say merely their names, but not even their exact shape and distinguishing marks. I shall say, "What in the name o wonder are they?" After a little looking at the strang pair I should probably say, "One is a kind of bear and the other is a kind of duck—a funny bear and a funny duck."

Explanation continued.—Observe how the process of interpretation of my impressions goes on. Looking at the Tasmanian Devil, my impressions divide themselves into two classes: one set of impressions resembles impressions of bears which I have previously received, while the other set finds nothing already existing in the mind to which it can attach itself. A kind of fight goes on between new and old. In the end the points of resemblance overpower the points of dissimilarity, and I judge the one animal (the T. D.), in spite of much unlikeness, to be a kind of bear, in doing which I am wrong, as it is a kind of marsupial, and in judging, by a similar process, the Ornithorhynchus to be a sort of bird, because of its bill, the mind equally makes a mistake, or, as we say, receives a wrong impression.

Two Points to be Noted.—There are then at least two parts in the process of knowing any object. First of all there is the excitation of our nerves, the nervous stimulus which makes us feel that we have a feeling, but does not explain what the feeling is, and secondly there is the interpretation of the feeling by a mental action through which the undetermined and as yet unknown sensations or gifts of the senses are referred to known impressions and explained.

III.

Assimilation of Impressions.—It is this act of mental assimilation of the impressions which we receive from external objects that I wish to discuss to-day. I am not dealing with the question of the origin of our impressions or the physiological basis of them, but with the growth of knowledge in the understanding by the working of the mind upon impressions. I think that modern psychologists have carried the analysis of this process sufficiently far for the results of their studies to be of practical value to teachers and parents. If we have to teach, is it not useful to know how the mind acquires knowledge? ·

An Example.—Take an object and set it before a child—say a fern. If the child has never seen a fern before, he knows not what it is. Impressions of it he receives, but he cannot interpret them adequately. The botanist looks at the same fern, and not only sees and knows that it is a fern, but also what kind it is, how it is distinguished from other ferns, where it grows, how it may be cultivated, and all about it. The difference between the knowledge which the sight of the fern gives to the child and to the botanist does not depend upon the fern, but upon the state of mind of the two observers. The mind adds infinitely more to the impression received when it is the botanist's mind which receives it, than when it is the comparatively empty and uninformed mind of the child. What you can know of an object depends upon what you already know both of it and of other things. Philosophers and poets like Kingsley, Carlyle, Herder, Goethe, as well as educationists and psychologists, impress on us this truth: "In regarding an object we can only see what we have been trained to see." *

Another Example.—Impressions, then, have to be *interpreted* before they are clear to us. What is the easiest case of our interpreting impressions? Perhaps some such as the following: I see a man a little way off and say to myself, "Here comes my brother." I have so often recognized my brother that the whole process of recognition goes on in my mind without any check or hindrance. The existing mental conception of my brother masters completely and promptly the fresh impressions which his present appearance makes upon me. The identification of the new and the old is uninterrupted, prompt, and immediate. The same speed and accuracy of interpretation is observable in his prompt

---

* CARLYLE.—We can only see what we have been trained to see.

GOETHE,—We only hear what we know.

HERDER.—What we are not we can neither know nor feel.

ROUSSEAU.—We can neither know, nor touch, nor see, except as we have learned.

In other words, the present impression produces only such an effect on the mind as the past history of the mind renders possible.

and correct recognition by a good reader of the words and sentences in his book.

A Difficult Example.—Now take an opposite case, when it is hard instead of easy to interpret impressions Suppose that we see something which is quite new to us. Suppose that the new impressions do not connect themselves with any previously assimilated impressions, and that try as we may to refer them to something known, all is in vain. Then we feel puzzled: a hindrance, or check, or obstruction occurs in our minds. If the impression be very strong, it may cause us to "lose our heads," as we say, or it may even overwhelm us.

The Impression may be " too much " for us.—It is narrated that one of the natives of the interior of Africa who was accompanying Livingstone to Europe no sooner found himself on the great Indian Ocean with nothing but heaving waters far and near in his view, than he became overpowered by the immense impression which this new experience made upon his mind, and flung himself overboard into the waves, never to rise again. Similarly at the Paris Exhibition, every evening when the gun is fired at the Eiffel Tower for the last time at ten o'clock, it is not unusual to see a sort of frenzy among the visitors. Under the already strong impression produced by the electric illuminations, the luminous fountains, and the varied magnificence of the great show, some people seem to be seized with a veritable panic. Cries of admiration escape from some, and terror from others, followed by fainting, attacks of hysteria, and prostration.

Another Case.—Similar shocks occasionally prove fatal. Only in September last, a little girl, four years old, was standing on the platform, near Sittingbourne, with her parents, who were on their way to Kent for the hop-picking season, when an express train dashed through the station. The little one was terror-stricken, and on the journey down screamed every time an engine came within sight or hearing. She dropped dead. The doctor ascribed death to the shock.

To assimilate then a wholly new impression is necessarily a task of some difficulty, but the results are luckily not always so sensational as those which I have just described, and the following is an account of what more usually takes place.

What Becomes of the New Impression.—If the new impression is not of a nature to make us feel strongly, and if it is isolated and unconnected with any other knowledge present to our minds, it probably passes away quickly and sinks into oblivion, just as a little child may take notice of a shooting-star on a summer night, and after wondering for a moment thinks of it no more; if, however, our feelings are excited, and if the object which gives the impression remains before us long enough to make the impression strong, then the impression becomes associated with the feelings and the will comes into play, in consequence of which we determine to remember the new impression, and to seek an explanation of it. With this object the mind searches its previous stock of ideas more particularly, comparing the new with the old, rejecting the totally unlike and retaining the like or most like, and in the end it overcomes the obstacle to assimilation and find a place for the new along with the old mental store thereby enriching itself, consciously or unconsciously—unconsciously in earlier years, and consciously after wards.

The White Violet.—As an instance, I will suppose child who has only seen blue violets finds a white on Of his impressions of the white flower, some are ne and some are old. The greater part are old, and lea him to infer that he sees a violet; but the impression o whiteness is new, and leads him to say, "This is not violet." Let us represent the characteristics by whic he recognizes a blue violet by the letters A B C D, th D standing for the color blue and A B C for all th rest of the flower. When now he finds a white violet h again notes A B C as before, but instead of D, the colo blue, he receives the impression E, the color white. Had the color been the same, the impression of the flower would have coincided with previous impressions of violets, but the difference between D and E causes an obstruction or hindrance to this inference. The mind is not at ease with itself; the agreement of new and old only reaches a certain way. The old mental image and the newly acquired one don't exactly tally.

White Violet (continued).—What happens? In the two mental images now present and side by side in the mind, the new and the old (the new being more vivid, the old being more firmly established), the like elements, namely, A B C, strengthen each other and unite to make a clear image, while the unlike elements D and E, the blue and the white, obstruct each other, become dim and at last obscured. The like elements in the end overcome the obstruction caused by the unlike and beat them out of the field of mental vision, so that the two partly resembling impressions become blended or fused, as by mental smelting, into one. The two are recognized as one by the mind. The old appropriates or assimilates the new. The child finds an old *Ex*pression for the new *Im*pression, and says to itself, " It is a violet."

The New Impression may Join at Several Points.—Of course an impression need not belong to only one previously acquired impression or group of impressions; it may be connected with other groups. In this case it will be recalled to consciousness on more frequent occasions than if it belonged to one other mental state only. Hence a new impression, *if you give it time,* may find for itself many more points of attachment with previous impressions and ideas than it found just at first.

Example.—For instance, I may visit Amiens Cathedral. Presently when I have admired the building I recall to mind various historic events that took place at the capital of Picardy. I remember that Julius Cæsar started thence to conquer Britain, that Peter the Hermit was born there, and that not far off Edward III. won the battle of Crecy, and that its name often comes up in the long hundred years' war. I think of the Peace of Amiens in 1802, the visit of Bonaparte to Amiens when he prepared to invade England, lastly of the German army in 1870. One impression calls up another, and the whole mass together strengthen and confirm and amplify the original impression. Isolated, these separate events are of less interest than when grouped together with my actual inspection of the ancient building.

Why we rightly Delay in Deciding.—A wise man, therefore (if I may draw a passing moral), does not, if he can help it, decide or act in a hurry, under the influence of new impressions, but he will give them time to find points of connections with old impressions. What may to-day seem irreconcilable with truth, or honor, or happiness, may prove when time has been allowed for assimilation inconsistent neither with sincerity, nor good name, nor good fortune.

The Plea of the Educator.—Educationists, like Mr. Arnold, also, will continue to implore the public to simplify the studies of children, being convinced that unless the mind has leisure to work by itself on the stuff or matter which is prescribed to it by the teacher, the thinking faculty, on which all progress depends, will be paralyzed, and dead knowledge will be a substitute for living. The mind will have no power of expanding from within, for it will become a passive recipient of knowledge, only able to discharge again what has been stuffed into it, and quite powerless to make fresh combinations and discoveries. Cram is the rapid acquisition of a great deal of knowledge. Learning so acquired, though useful for a barrister, has less educational value than the public believe, for it does not promote but rather tends to destroy the active and constructive powers of the mind.

Effect of the Assimilating Effect.—When the mind has much difficulty to overcome in assimilating a new impression, and hence has to spend time in so doing, it is benefited by the process, for in the first place the necessity of care, caution, and accurate observation, and much rummaging (if I may venture on the expression) among the ideas in the mind tend to sharpen the senses, the sight, the touch, the hearing, and the rest, by making them sensitive to fine shades which might otherwise escape us, and in the second to amplify and enlarge meagre impressions.

How we know Solids.—The eye, by itself, for example, only reveals to us surfaces. How then do we seem to see solid bodies? A baby stretches out its hand for the moon: how is it that what seems so near to him looks so far off from us? Because in our case the impressions conveyed by the eye are supplemented by the impressions received through the touch, and the two distinct sets of impressions combined together in the mind furnish us with the conception of a third dimension, besides length and breadth—viz., depth. The child who has not yet got so far as to have sufficiently often united the impressions derived from looking, with those derived from touching and moving cannot rightly interpret the impressions which he receives. The moon seems quite close to him.

Unexamined Impressions do not yield Clearness.—Impressions, on the other hand, which pass easily into their place in the mind do not always tend to clearness of ideas. People may look at an object hundreds of times for a special purpose, and beyond serving that purpose get no permanent impressions at all. Many people who look at a clock or watch many times a day cannot at once, when asked, draw from memory a dial with the hours correctly placed upon it.

Why Assimilation may Mislead.—The process of assimilation may even mislead just as familiarity with an object may hinder accurate observation. Goethe says there is a moment in his life when a young man can see no blemish in the lady he loves, and no fault in the author he admires. A man in love may think that his Angelina sings divinely sweet, though her voice is like a crow's. He interprets the impressions which he receives according to previously formed impressions.

Fault not with Senses.—This leads us to see that it is not right to say, as we sometimes do say, "My senses play me false." The senses do not lie. The ear does not in the instance in question convey sweet sounds. The sense of hearing does not judge at all. The ear conveys the sound truly enough. The judgment concerning the sound is made in the mind of the listener. This judgment it is which is falsified by prejudice, the lover being naturally prepossessed in favor of his mistress.

So the wanderer in the graveyard by night in the uncertain light of the misty moon judges a tall gravestone to be a "sheeted ghost." His eye is not at fault. His judgment is. He receives the impression from the object truly, but he refers his impression to the wrong group or store of previous knowledge. He should refer it to optical phenomena, diffraction of light and the rest. He actually does think of pictures and stories of vague appearances of human shapes without human substance, and all the superstitious imaginings of **poor**

frail human nature. His senses are not under control
of his reason.

The Perceptive Process modifies Previous Knowledge.
—We have seen then how each impression that we receive from external objects is consciously or unconsciously interpreted and made known to us by a kind of
internal classification through which it is referred to that
part of our store of knowledge to which its resemblance
connects it. We have now to see that in this process
of interpretation of a new impression by that which is
old, the previously existing mass of knowledge which interprets the new is itself modified and made clearer.

Example.—Suppose a child lives in the flat of the fen
near Cambridge, and that by going to the Gogmagog
Hills he learns to form an idea of what a hill is. Then
suppose him to be transported to Birmingham, where
he goes out to the Lickey Hills. These he will recognize as hills by aid of the previous conception of a hill
which he has formed in his mind, but at the same time
he enlarges his ideas of a hill, and if he travels further
west and climbs the Malvern Hills and the Welsh Hills
he will still further amplify his conception. Now let
him study the elements of geology and physical geography, and learn to trace the connection between the
shape of hills and the rock or soil composing them,
together with the action of wind and water, heat and
frost, and the word " hills " will still have yet an extended
meaning.

Still Further.—Every time you refer an object to a
class, as when you say, " Yonder mass—it may be Ingleboro'—is a hill," you not only explain the thing about

which you are talking (Ingleboro'), but you also add to
your idea of the *class* to which you refer it (hill). The
new thing is explained by old or already existing ideas,
but for the service which the old does the new in thus
interpreting it the old idea receives payment or recom-
pense in being made itself more clear.

Suppose you have a dozen pictures, apes, bears, foxes,
lions, tigers, etc. Then every time you show one of
these to a child, and the child learns to say "that tiger
is an animal," "that lion is an animal," he not only
learns something about the tiger, the lion, and the rest,
but also extends his conception of what an animal is.
Hence we can see when it is that learning a name is
instructive: it is when the name is a record of something
actually witnessed. If, however, you tell a child who
does not know what a ship is, or what wind is, or what
the sea is, that a sail is the canvas on which the wind
blows to move the ship across the sea, the names are
only names, and do not add to his knowledge of objects.

The Interpreting Process continually at Work.—So
far we have chiefly considered the case where impressions from the outside world or from *outward objects*
are being interpreted by the mind, as in the case of
violets, the pot of ferns, and the like; but a similar
process goes on wholly in the mind between ideas which
exist there after external objects have been removed.
Consider how weak fugitive impressions may be
strengthened and held fast by this process. Alongside
the feeble, and therefore fugitive, impression arises a
mass of previously acquired and nearly connected impressions and ideas, dominating the former, and by

means of connections with other stores of knowledge
setting up a movement in the mind which lights up the
obscure impression, defines it, and fixes it in the mind
ineradicably.

Example.—For example: I find a little white flower
on the top of Great Whernside, *Rubus Chamæmorus.*
I might notice it for a moment and pass on oblivious.
Suppose, however, that it occurs to me next day to
think of the so-called zones of vegetation, and how the
Pennine Hills were once covered with the ice sheet like
Greenland now is, and how England then had an arctic
flora, and how it may be that this flower, which in England only grows 2000 feet above the sea, being killed by
the warmth of lower levels, may perhaps be a botanical
relic of that surprising geological epoch, and then what
interest attaches to that flower. Why the very spot on
which it stands seems stamped in the mind indelibly.

The Psychological Process by which we Learn.—
Nothing new then can be a subject of knowledge until
it is not merely mechanically associated (as a passing
breeze with the story which I read under a tree), but
associated by a psychological process, with something in
the mind which is already stored up there, the new
seeking among the old for something resembling itself
and not allowing the mind peace until such has been
found, or until the new impression has passed out of
consciousness.

Not Necessarily a Recognition of Self.—This process
of interpreting impressions and ideas by reference to
previous impressions and ideas must not be confounded
with the reference of such interpreted impressions to

self. When you refer this process to self, when you recognize yourself as going through the process, and as being the subject of the assimilating process, this is self-observation. You may have this self-consciousness either along with the interpreting process, or after it, or not at all. Dogs, parrots, and many animals clearly interpret impressions and objects as one of a class,—as a kitten did who, after eating a piece of raw meat, afterwards chewed a ball of red blotting-paper, inferring it to be meat from its color; but they do not do this with recognition of self as the subject of the process. Children do not appear to be conscious in their thoughts and actions much before they are three years old, and their minds seem at first much to resemble the minds of animals.

Application of the Principle in the School-room.—We may now further apply this principle of the growth of the mind to practical work in the class-room. When something new presents itself to us, it does not, as a rule, except when it affects the emotions in some way, arrest our attention, unless it is connected with something already known by us.

Example.—A young child visited the British Museum, and was next day asked what he had noticed. He remarked upon the enormous size of the door mats. Most other impressions were fugitive, being isolated in his mind. The mats he knew about, because he compared them with the door mat at home. Among all the birds, the "only one he remembered was the hen, and passing by the bears and tigers with indifference he was pleased to recognize a stuffed specimen of the domestic cat. The child only remembered what he was already familiar with, for the many impressions from other objects neutralized each other, and passed into oblivion.

Apply the Principle.—One great art in teaching is the art of finding links and connections between isolated facts, and of making the child see that what seems quite new is an extension of what is already in his mind. Few people would long remember the name and date of a single Chinese king picked by chance from a list extending back thousands of years. Facts of English history are not much easier to remember than this for children who are not gifted with strong mechanical memories. Hence the value of presenting names, dates, and events, in connection with external memorials, such as monuments, buildings, battlefields, or with poems and current events, and the like. Story, object, and poem illustrate and strengthen each other. It ought not to be hard to teach English history in the town of York, where there is a continuous series of objects illustrating the course of affairs from prehistoric times to the present date. Our object in teaching should be to present facts in organic relation to each other, instead of getting them learnt by heart as a list of disconnected names.

The Child's First Task.—If, then, all the growth of the mind takes place from earliest to latest years, through the apprehension of new knowledge by old, then the first business of the young child in the world is to learn to interpret rightly the impressions that he receives from objects. To receive and master the gifts of his senses is his first duty.

Does not Proceed Systematically.—But this task cannot in the early stages be fulfilled in a strictly systematic way. You cannot present all the world piecemeal to the child, object after object, in strictly logical order. One educationist objected to little children visiting a wood or forest because the different sorts of trees were there all jumbled together instead of being scientifically classified and arranged as they would be in a botanical garden. The child, however, must take the world as he finds it. Impressions come crowding in upon him in such numbers that he has no time at first for paying minute attention to any one. In truth so massed and grouped are his impressions, that one may almost say that the outer world presents itself to him as a whole—of course an obscure unanalyzed whole,—and that it is a matter of difficulty to isolate one perception clearly from its concomitant perceptions.

Impressions from Actual Life Lead.—The whole must be analyzed into parts bit by bit. Out of the mass of obscure and ill-defined impressions, educationists should study which are they which stand out and arrest attention most readily, and in what order they do this? We do not find that those impressions are most striking which are logically the most important, but rather those to which the practical needs of daily life give prominence—food, clothing, parents, brothers, sisters, other children and their experiences. Such are the things that children are most taken up with. But each impression once grasped is the basis or starting-point for understanding another, and thus the manifold variety of objects is simplified and brought within the compass of memory by a sort of unconscious reasoning.

Example.—A child, for instance, who kept a chicken, but never saw a chicken at table, being limited in its meat diet to beef, when at last the chicken came to table roasted, called it "hen-beef," clearly interpreting by an elementary process of reasoning the new by the old. Take a child to a wild-beast show and observe how he names the animals by aid of a very general resemblance to those he may previously know. The elephant is a donkey because he has four legs; the otter is a fish; and so on. These comparisons are not jests, nor even mere play of fancy, but the result of an effort of an inexperienced mind to assimilate new impressions. The child is only following the mental process which we all have to follow in becoming masters of our impressions and extending our knowledge. Clearly the limited stock of ideas of the child renders it easier for him to make mistakes than for us to do so, but in some matters it is well to remember that we are no further advanced than children, and consequently often behave as such.

Another Example.—A little French child, a year old, who had travelled much, named an engine Fafer (its way of saying *Chemin de fer*); afterwards it named steamboat, coffee-pot, and spirit-lamp, anything in short that hissed and smoked, "fafer"—the obvious points of resemblance spontaneously fusing together in the child's mind and becoming classified, not quite incorrectly. Another child who learnt to call a star by its right name applied star as a name to candle, gas, and other bright objects, clearly interpreting the new by the old, by use of an unconscious elementary classification or reasoning.

Value of Names.—Thus we see the value and helpfulness of language, in the process of acquiring and interpreting impressions. Having once separated out from the indistinct masses of impressions borne in upon him from the outside world some one distinct impression, and having marked that impression with a name, the child is thenceforth readily able to recognize the same impression (in this instance that of brightness) when mixed up with quite other masses of impressions, and to fix its attention on that one alone.

The Word a Mental Help.—Thus the word helps the mind to grow and expand. The use of the word is a real help to the knowledge of things. The name when learnt in connection with the observation and handling of an object is not merely a name, a barren symbol for nothing signified, but is a means for acquiring fresh knowledge as occasion serves. A name thus learnt (i.e., in presence of the object), when applied by the learner to a new impression exactly resembling the former, is really an expression of and an addition to the mental stores. It is then as the filling in of a sketch or as the further completion of an unfinished circle.

Names should be Given in Presence of Object.—How different is such naming from learning by heart of names of objects without handling the things signified. How often have text-books of science, geography, and history been prescribed to be got up for examination,

and how often have the results been disappointing. The
student thus taught sees only the difference of a letter
in the alphabet between Carboxic Acid and Carbolic
Acid, Jacobix and Jacobite, and a mere transposition of
a figure in expressing an incline as 8 inches in 1 mile,
instead of 1 inch in 8 miles. The words call up no
mental image. The figure 8 is a symbol only as it does
not call up the image of 8 things. A name given in the
presence of the object serves afterwards to recall the
image or picture of that object, and it does this the
more perfectly the more accurately the object is studied
in the first instance.

Why Gestures are Used.—Children for want of language signify many of their impressions by gestures
before they can describe them in words; and *gesture*
language, especially if encouraged, precedes spoken
language, besides accompanying it. Children are imita-
tive; they love to act over again what they have seen,
especially when much impressed, as in George Eliot's
pathetic description of the baby-boy attending his
mother's funeral in puzzled wonder, and thinking how
" he would play at this with his sister when he got
home." With children, this " acting," or "playing at
being," more resembles talking over, giving expression
to and describing what has been seen, noted, and assimilated, than aimless exercise of the muscles and the
intelligence.

Fröbel.—How profoundly right, therefore, Fröbel
was in making so much of action-songs in his Kindergarten, and how excellent his games are in which every

action of the child corresponds to some observed impressions with which the child is familiar.   Fröbel's actions correspond to realities, and are not mere physical movements.   They are forms of expression of things.   They correspond to facts, and advance the observation and knowledge of things which ought to be familiar to every one, such as sowing, reaping.

IV

Why the Child said " Feathers."—Now to go back to
my pot of ferns. The child sees ferns for the first time,
and cannot tell what they are. He receives impressions
which are new, and these seek interpretation in the man-
ner which I have described. They hunt about in the
mind for similar impressions previously received; at
last the impression of the fern attaches itself to the
impression of feathers; the crisp curl of the frond
and its delicate branches much resemble feathers; it
is true there is a hindrance to the judgment; the
fern is not quite like the feather; some points are like
and some are not; in the end, however, those which are
alike overpower those which are unlike, and the child
says, " These ARE feathers."

Such a Reply not to be Censured.—The child has not
got false impressions; he interprets wrongly; further
study, fresh observation and comparison, will soon rec-
tify the error. Hence the need of taking careful note
of children's mistakes, distinguishing between thought-
less answers and those which, although very wrong,
arise from mental effort misdirected. Careless answers
should be checked, but well-meant thought, even if un-
successful, should be encouraged. Therefore an answer

like that of the " green feathers " should be dealt with
in the way of praise rather than censure.

Effects of the New Ideas.—Sometimes it is not merely
an object that is incorrectly interpreted, and subse-
quently better understood. It occasionally happens to
us that a whole group of thoughts is thus modified by
the acquisition of some new knowledge, and instead of
the new merely forming an addition to the old, it wholly
changes it. Such was the result of the teaching of Co-
pernicus and Galileo, and in our own day of Darwin.
The discoveries of these men caused such wide-reaching
alteration of preconceived ideas that the new knowledge
was at first received with discomfort and mental un-
easiness, which caused the discoverer to be looked upon
with suspicion, regarded as any enemy, and persecuted.
When in the case of an individual some new conception
changes the character in this way by some powerful influence, as in the case of St. Paul, we call it a " conver-
sion."

The Old Finds a Place for the New.—Well, then, it
may be said, in these cases your position is given up.
The new should be regarded as the means by which the
old is known, instead of the old as interpreting the
new. But this is not the case, for however overpower-
ing the new conception may be for a time, yet in the end
the whole store of knowledge in the mind proves too
strong for it, overpowers it, and finds some place for it,
after which the mind is at peace with itself, and appears
to have been enlarged and not diminished or divided by
the fresh experience, however strange and unusual it
may have been.

Children need Aid in Naming.—I have shown, then, that when the child called a pot of ferns a " pot of green feathers " he was by no means using a name without attaching any meaning to it, and that he should have been encouraged for a praiseworthy effort to explain what he saw. It is, however, the business of parents and teachers to help the child to learn exactly what it is that he names.

Example.—A child, for instance, saw a duck on the water, and was taught to call it " Quack." But the child included in this name the water as well as the duck, and then applied it to all birds on the one hand and all liquids on the other, calling a French coin with the eagle on it a " Quack," and also a bottle of French wine " Quack." Such a mistake in naming is to be guarded against, as obviously tending to confusion of thought. The poet Schiller as a child lived by the Necker, and called all rivers which he saw " Necker." Such an error is less serious as it is easily put right. If the child notes its impressions and refer them intelligently to previous impressions as best it can, then it is not important if he is not quite correct about names.

V.

The Lesson for Teachers.—We—teachers and parents —may take a hint from this, and be more ready to give class-names to begin with, leaving details to come later. Teach the child in front of a picture of a herring, or better, pictures of herring, sole, and pike, "That is a fish" first of all, and only afterwards "That fish is a herring." For teaching general names, such as bird, beast, fish, and reptile, in presence of pictures of eagle, cow, herring, and adder, has a twofold use. The class name (fish, beast, etc.) thus given (1) directs the child's attention to a few points among many, and those easy to grasp, and hence is a guide to the child's mental powers, which are apt to be overwhelmed by the number of individual impressions of things, all disconnected and isolated, much in the same way as in an intricate country full of cross-roads your way is made easy if you are told to ignore all other tracks and follow the road bordered by telegraph-posts, and (2) it enables the child to understand the usual conversation of its elders and the words and language in books.

Relation of Names to Knowledge.—Grown-up people use general terms in daily conversation which children only slowly acquire without help from teachers. Many of these simpler class names are easily taught and are a pleasure to the children to learn, for they answer to the natural early stages of elementary reasoning. Country children often have a small vocabulary of general terms compared with town children, and less understand the language of books; but on the other hand, from exercising their senses on objects and being brought into close contact with out-of-door work they often have a greater real power of observing and interpreting things outside themselves, and greater originality in this respect than town children, who are sharper in talk and society. However, both kinds, the knowledge of language and the mastery of objects, should be taught together, for both are indispensable in life.

Children note Superficial Resemblance.—Young children are perhaps quicker than older people to note superficial resemblance of things. Because, no doubt, they have fewer old impressions stored in the mind wherewith to compare new impressions, and comparison among a few things is more rapidly and expeditiously made. They have to pay for this advantage, however, because they are liable to misinterpret impressions—to call a pot of ferns a pot of feathers, to refer impressions to the wrong group in their mind, groups with which they are accidentally and not logically connected.

Often Endow Objects with Personality.—What is more, objects are not so clearly distinguished—set over against each other—with children as with grown people. Children hardly distinguish themselves into soul and body. They know of their undivided personality—body, mind, and soul—that it moves, feels happy, sad, hungry, etc., and they attribute the same feelings to all other things.

Birds, beasts, and inanimate objects are like affected as themselves. " Jack the dog is thirsty," " Poll is angry," "Kitty is sleepy," " the stars blink," " the engine goes to bed," " the knife is naughty to cut me." They do not distinguish between figures of speech or metaphors and realities. Their minds move in a region of twilight, in which the real and the unreal, the true additions to knowledge, the actual gifts of the senses, are confused and blurred and altered by the additions which the mind itself makes to them, and they cannot separate the one from the other.

Hence the Use of Fairy Stories.—To this stage of mental progress how appropriate are fables, allegories, fairy stories, parables, and the like. If any one thinks that it would be better if the child's mind could move only in the sphere of the exact, I would reply, (1) that this does not seem to be nature's process, (2) that looking to the mode of growth of the mind it does not seem even possible, and (3) that if you try to keep the child's mind to exactness you may clip and pluck the wings of imagination.

Imagination Important.—Now without imagination there is little advance in knowledge, little discovery in the sphere of science and in the sphere of morality; without some imagination you are quite unable to put yourself in the place of another, which is the basis of sympathy and mental support, and the foundation of the social fabric. The mere sight of a neighbor's joy or sorrow does not awaken sympathy.

Example.—Three little children were thrown out of a train in an accident, and one was frightfully mangled to death; but the other two, who were unhurt, and could not realize what had happened, stooped down and went on plucking daisies with unconcern. In the case of young children you can hardly go too far in the way of associating new learning with personal feeling, even at the expense of exactness, and the infant-school teacher who, in a lesson on the sun, instead of dwelling on its roundness, brightness, and heat, began by calling it a lamp in the sky, lighted in the morning and put out at night; lighted for men to go about their work, and put out for them to go to sleep, showed a true knowledge of the key that opens the door into the child's mind.

Childrens' Ways must be Studied.—This information is not exact, but inasmuch as it is based on what children understand and like to hear about, it finds a ready entrance into their minds. But it is clear that what is to the child its natural mode of expression is arrived at by the teacher only through imagination, and hence arises the teacher's difficulty. It is a useful hint to study the children's own lead and follow it. School necessarily limits the child's life. You cannot bring all creation into the four walls of the class-room. But what you lose in extent you gain in depth: you lose variety, you gain in concentration. Before school-time, all things engage the child's attention in turns, and nothing long. At school he has to attend to a few things, and to keep his attention fixed upon them for short periods at first, but for increasingly longer ones. It is a matter of practice and experience to find out what things most readily arrest attention, and in what way information can best be conveyed so as to arrest attention, and it is in these matters that the skill of the teacher comes in.

The Teacher's Art.—I am not sure that if the teacher's art is to be summed up briefly it may not be described as the art of developing the power of fixing attention. For instance, when we present a picture or even an object to a child, neither object nor (still less) picture explains itself. The object needs to be pointed out piecemeal, and all its parts called attention to separately, for the child only sees it as a whole, about which it can say but little and soon tires of. The picture but very partially represents the objects which the artist depicts, much being suggested and left to the imagination of the beholder. Even when we say we actually see an object we forget how much of what we think we see is really inference from some small part of what we see, and nothing is more deceptive than merely ocular evidence. Thus pictures of things which the children have seen are much better to commence with than pictures of things which they have not seen, and the former should serve as a preparation for the latter.

Difficulties in Teaching.—But even pictures will only go a certain way in making known to us things past and things remote, facts of history and geography. The greater part of advanced instruction must be conveyed by words. Is it an historical scene we are treating of? The child and many grown people interpret all by their own experience; towns and houses in history resemble in his mind those with which he is familiar; men and women move about in the dresses of his near neighbors; their aspect and language are in his mind the same as those of his people with whom he daily converses. Such inaccuracies may be partly corrected, but in the main they are unavoidable. History cannot be communicated with complete truth; the lives of men and women personally unknown can be only partially conceived. Hence Goethe says, "The past is a book with seven seals."

The Past Known through the Present.—The best plan is to read the past with one eye on the present. Look at the pictures of the Holy Family as drawn by Italian and Dutch painters. The chief fact which they intended to depict is not obscured but made clearer by the painter having made the homely surroundings French and Italian rather than original. In History and Geography, in order to help the child to understand old times and realize what distant lands are, we must store his mind with conceptions based upon frequent observations of present time, and of his own home and its surroundings.

Example.—How far such observations may carry the student in interpreting the unseen, is proved by the beauty and correctness of the descriptions of Alpine Countries, which were written by Schiller before he had seen the Alps. In history the most human part of the narrative takes the firmest hold of the mind, and the story of "King Alfred and the Cakes," though not a very noble historical anecdote, serves at least to fix the name of the king in the child's mind, who would not so easily remember the peace of Wedmore. Eating he knows more about than making treaties.

VI.

Advanced Stages.—We may now trace the process of acquiring knowledge in its more advanced stage. The child has now learned that a pot of ferns is not a pot of feathers. Perhaps, however, he has only seen one kind of fern—say a Lady-fern. After a few weeks he may see another—perhaps a Maiden-hair. The points of resemblance between the two make him say, "That is a fern:" the points of difference hinder the process of assimilation and make him doubt; in the end the mass of old impressions resembling each other over-power impressions which differ, and he says, "This is a fern," and in so doing he enlarges his conception of what a fern is.

Let us now suppose that he comes across a good teacher who shows him many kinds of ferns, and points out the difference between ferns and flowering plants and mosses. Every fresh distinction, every observation of a new fern, helps to modify his previous knowledge. Old and new impressions react on each other.

Now Uses Judgment.—But now mark how essentially the same and yet how different are the two mental states: the earlier one, namely, when the child (I would say the child's mind) recognizes of its own accord the second plant as a fern by means of its previous acquaintance with another fern, judging from a more or less superficial resemblance, and the latter state of mind when he has learnt all the scientific distinctions by which a fern is classified in a different class from flowering plants and mosses. We have now passed from Infant School learning to the instruction which is appropriate to the Upper School and the advanced classes. The child has outgrown a state in which the mind reasons unconsciously, and has arrived at a state in which reasoning is conscious; he has left behind a condition or stage of development in which he was at the mercy of his impressions, and has progressed to a state of mind in which he can compare, check, and control his impressions. He has passed from a state in which he unconsciously accepted what was present to his mind to a state in which he can infer, judge, and criticise.

Looks More Closely.—The pot of ferns is now seen to have more points in which it is unlike feathers than points in which it resembles them. Of the many impressions derived from looking at the pot of ferns, the feather-like impression which at first stands out from the rest and forces itself on the mind, to the exclusion of the other impressions which would, if attended to, modify the judgment, is now by means of conscious reasoning brought under proper control, and put in a subordinate position. What appeared to be a fact is now seen to be a fancy, and after all a fancy which expresses some element of truth—viz., the resemblance between ferns and feathers.

As Illusions and Fancies.—These considerations, perhaps, throw some light upon Dr. Allbutt's warning to

parents about the dreams and illusions of children.
The fancies of childhood, he thinks, are sometimes the
ante-chamber of insanity in adults. I do not think he
intended to knock on the head many poetic and popular
conceptions about children's pretty fancies, as was stated
in some evening review of his remarks.* It is clear,

---

* CHILDHOOD'S DREAMS: IMAGINATION OR INSANITY? In the
course of the meeting of the Medico-Psychological Association
held at York Dr. Clifford Allbutt (of Leeds) read a paper on the
"Insanity of Children," which, if its statements be well founded,
knocks on the head many poetic and popular conceptions. Words-
worth speaks of a child's ideas being a reminiscence of "the fairy
palace whence he comes." Dr. Allbutt sees in them only a step
towards the insane asylum. Most people regard it as a healthy
sign if the children have pretty fancies, and those are thought to
be happiest who keep their illusions longest. But Dr. Albutt
would reverse this judgment. The fairy dreams of childhood are
only the result of defective organization, and healthy growth con-
sists in their evaporation.

Here are some of the chief passages in Dr. Allbutt's paper :
The insanity of children was the vestibule of the insanity of
adults ; in children they saw in simple primary forms that with
which they were familiar in the more complex and derivative
forms of insanity in adults. If a man lived in a vain show, far
more so did the child ; if a man's mind was but a phantom in relation to the world, so fantastic was the child's mind in relation
to that of the man. Fantastic—that was the key to the childish
mind. In him was no definite boundary between the real and the
unreal. Day-dreams which in an adult would be absurd, were to
a child the only realities. As the child grew older, and sense
impressions organized themselves more definitely and submitted
to comparison, fantasy became make-believe and the child
slipped backwards and forwards between unconscious, semi-con-
scious, and conscious self-deception. Pretty were the fancies of a

however, that the crude method of assimilating knowledge, which is natural and apparently inevitable in a child, ought by degrees to yield to more accurate conceptions under the influence of wise instructions.

Suggestions.—It is one thing to confuse ideas unconsciously; it is another thing to do so consciously. The child makes an unconscious mistake in calling ferns feathers, but if this confusion is cherished by the child after he well knows the real distinction between the two, and if he acquires or cultivates a habit of mind in which reality is made to give way to make-believe and pretence, the child may lose control over its judgment and become in the end imbecile. The best antidote to foolish imaginings appears to me to be the time-honored fables of Æsop, the sacred parables and allegories, and the best modern fancies for children, like those of Andersen or Ruskin. Fantastic the child will be it is our business to make his fancy healthy.

The Object of Learning.—The object, then, of learning in education is not only to make the mind fuller and to enrich the understanding, but if the instruction be of the right kind the additional knowledge ought to

---

child, yet the healthy growth of the child consisted in their evaporation. But if the growth of the mind were something other than healthy, then these fancies kept their empire ; they did not attenuate, and the child did not put off its visions. They were not likely to forget that the persistence of insanity in children might prevent the due advance of the organization of the results of impressions, and might ultimately, as the adolescence approached, leave the sufferer in a state of more or less imbecility.
—*Pall Mall Gazette.*

make the old knowledge more exact and better defined. The method of acquiring the extended knowledge, also, ought to have even more far-reaching results than the information itself. Accustomed to right methods of study the child will learn to be cautious in dealing with fresh impressions, to feel the pleasure of receiving new impressions and the need of care in referring them to their proper class, to realize the danger to which every one is liable of forming hasty judgments, and to weigh evidence for and against a provisional judgment.

Further.—In short, study ought at least to make the student acquainted with the limits of knowledge in general, and the limitations of *his* knowledge in particular. The country proverb, " He does not know a hawk from a heronshaw," illustrates the sort of progress that learning should produce in a child. He must acquire at school the power of apprehending quickly and correctly. He must become sharp in receiving impressions, and accurate in referring them to the class to which, not fancy, but reasoned judgment, leads him to refer them.

How Educate by Acquiring Knowledge.—Accurate and complete conceptions, true logical definitions in *all* matters that we deal with in daily life, cannot be obtained by any of us. We can only keep the ideal of perfect knowledge before our eyes as a guide to us in the path of right knowledge. The educational value of the acquisition of knowledge is to improve the natural powers of thought and judgment, and to enable the learner to deal with the masses of observed facts which press more and more heavily on us as we have to

move amid the complications of mature life. In acquiring knowledge the mind is naturally active, and not
merely passive. The active element is most precious,
and modern education often tends to strangle it. Yet
instruction which does not add increased energy to the
thinking powers is failing its purpose. Learning cannot be free from drudgery, and a great deal of the process of teaching and learning—say what you will—must
be a tax on patience and endurance; neither can we
entirely dispense with the mere mechanical exercise of
the memory; but if the method pursued is correct, the
drudgery ends in an increase of the energy of the mind,
and a desire and a power to advance to new knowledge
and discovery.

Two Purposes in Education.—You cannot undertake
at school to fit every child for entering a trade, or craft,
or profession, without further learning; but what he has
learnt as a child ought to develop his constructive facul-
ties, and to enable him to deal effectively with the matter which he will have to handle in the stern school of
life. And if, in addition to this, he has acquired an in-
grained preference for the good before the bad, the true
before the false, the beautiful before the foul, and what
is of God before what is of the Devil, his education has
been as complete as it admits of being made.

Can only develop Original Power.—As in the early
stages of life, so in the later, our knowledge and our
conduct depend as much on what is within us as on
what is without. The work of life cannot be well done
mechanically; in this every one must be partly original
and constructive, for the world is not merely what we

find it, but partly what we make it, and what Coleridge
has finely said of Nature applies to all we think and do.

> O Lady, we receive but what we give,
> And in our life alone does nature live;
> Ours is her wedding garment, ours her shroud !
> And would we ought behold of higher worth
> Than that inanimate cold world allowed
> To the poor, loveless, ever anxious crowd ?
> Ah ! from the soul itself must issue forth
> A light, a glory, a fair luminous cloud
>   Enveloping the earth.

That education is the best, not which imparts the greatest
amount of knowledge, but which develops the greatest
amount of mental force.

www.ingramcontent.com/pod-product-compliance
Lightning Source LLC
Chambersburg PA
CBHW022026190326
41519CB00010B/1618